A Symmetric Concurrent B-tree Algorithm

Vladimir Lanin,
Dennis Shasha

Courant Institute,
New York University

This work was partially supported by
the National Science Foundation
under grant DCR8501611
and by the Office of Naval Research
under grant N00014-85-K-0046

1

ABSTRACT   We present a method for concurrent B-tree manipulation in which insertions are performed as in an earlier paper by Lehman and Yao, and deletions are done in a symmetrical, novel fashion   The result is an algorithm that almost always holds locks on only one node at a time   To allow this low level of synchronization, the integrity constraints on the data structure are re-examined   This is useful in verifying the algorithm by a general semantic serializability proof method   Simulation shows that the algorithm is capable of achieving significantly better concurrency than other algorithms that perform insertions and deletions symmetrically

1. Introduction

B-trees are a popular implementation of the *dictionary*, an abstract data type (ADT) that supports the actions *search*, *insert*, and *delete*   A full review of B trees may be found in [Co79], but let us say here that a sequential algorithm for performing the above actions in a B⁻tree usually runs in three stages   a descent through the tree to a leaf node, an operation on the leaf node that either checks for the existence of a key adds a key, or removes a key (for search, insert, and delete, respectively), and an optional ascent during which the tree is restructured in order to rebalance it   It is also possible to restructure the tree during the descent, as described in [GuSe78]   Restructuring is done by *splitting* a single node into two neighboring ones and by *merging* neighboring nodes into one   The execution of an entire action takes time logarithmic in the number of keys stored in the structure

It is useful to allow several different, asynchronous processes to access the B-tree concurrently   This is easy when all the processes are searches, but the goal is to also allow concurrent insertion and deletion, while ensuring correctness and allowing as much concurrency as possible

1.1. Previous Work

The first concurrent B-tree algorithms by Samadi ([Sam76]) and Bayer and Schkolnick ([BS77]) ensure correctness by completely isolating each action from the effects of all other actions active in the structure concurrently a concurrent action never encounters a situation that it could not encounter if executing alone   Two techniques (reminiscent of [KS83] tree protocols) are used to achieve this isolation

a) An action descending from a parent node to one of its children must acquire a lock on the child before releasing the lock on the parent (lock-coupling)   Without this lock on the parent, a writer action could acquire locks on both nodes, change them (e g split the child), and release the locks (Fig 1)   This could cause the descending action to perform incorrectly, as the search(d) in fig 1 does

b) A writer action (which in [Sa76] and [BS77] restructures during the ascent) must lock out other writer actions from the entire subtree dominated by the highest node it has to modify   This averts another danger suppose writer w needs to ascend to and modify node n, and writer w' is already active below n Since w' might also have had to modify n, either w or w' would update n first   The other would then access a different version of n than during its descent, again causing errors in some cases

These techniques guarantee correctness at the expense of concurrency   Locking processes out of entire subtrees is a severe disadvantage in itself   Furthermore, the most straightforward method of achieving the subtree-locked state (which is used in the above algorithms) is for writer processes to begin placing exclusive (or at least writer-exclusion) locks from the top of the tree   This temporarily locks out new actions entirely   The

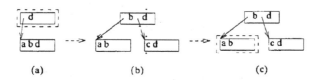

(a)   (b)   (c)

Figure 1   A sequence of events leading to error
in the absence of lock coupling

Descending action *search*(*d*) releases its lock (shown in dashes) on the parent node in (a)
before acquiring lock on child node in (c)   Meanwhile, action *insert*(*c*) splits the child node
in (b), making the key *d* unavailable to the search

simultaneous use of lock coupling by other processes aggravates the problem by increasing
the interference

Another algorithm in [BS77] mitigates the above effects of subtree locking by having
writers first perform an optimistic search like descent that uses read locks on all nodes but
the final leaf, on which it uses an exclusive lock   A writer can complete after the optimistic
descent if it does not have to split or merge the leaf   This has a probability of about $\frac{k-1}{k}$ if
the number of keys stored in each leaf ranges from k to $2k$ Otherwise, the writer must give
up and re-descend using the normal protocol

An algorithm by Mond and Raz [MR85] is a concurrent version of the [GuSe78]
algorithm, and avoids subtree-locking by doing restructuring during the descent   However, it
still employs lock-coupling and requires using exclusive locks starting at the root   We note
that the [MR85] algorithm can also be enhanced by optimistic descents

Simulation shows (as presented in this paper, as well as in [El80], [Sh84]) that in a high
update environment, algorithms that use exclusive locks high up in the tree in the in
descent can not achieve significant concurrency   Although optimistic descents
considerably (especially in the [MR85] version), they depend on the number of keys in a leaf
being large   They seem to employ more locking than absolutely necessary

1 2   The Lehman and Yao Algorithm and Deletions

A more radical approach to achieving a minimal amount of locking is not to try to
isolate actions from each other (by lock-coupling and subtree-locking), but to enable them to
recover from the effects of other actions   Ellis [El80], and Lehman and Yao [LY81] have
written such algorithms   The latter avoids both lock-coupling and subtree-locking and is
simple, we base our approach on it

The [LY81] algorithm introduces the *B-link* structure, obtained from a B⁻ tree by
connecting each level into a singly linked list   A pointer called a *rightlink* points from every
node to its right neighbor, so actions can stray to the left of the direct path to a desired node,
and still recover by following the rightlinks   For example, a split occurs in two stages (see

Fig. 2): first, half the data is moved out of n into n', and n' is inserted into the linked list; next, a pointer to n' is put into the parent of n. Between the two stages, descending processes can be allowed to make the transition from the parent of n to n, because even if they need data that was moved to n', they can still reach it. Thus, lock-coupling is unnecessary. Subtree locking is unnecessary because ascending writers can follow rightlinks to the appropriate place to make their updates.

The major problem with the [LY81] algorithm is that no provision is made for the merging of nodes. Thus, the structure never becomes smaller even after many deletions, although Lehman and Yao propose to rebalance it "off-line". This deficiency has been addressed in the algorithms of [Sag85] and [Sal85], which give procedures for restructuring the entire tree that can run concurrently with the other actions. Since such procedures must lock out at least other writers form the root of the tree, performing them too often can decrease concurrency. However, no direct connection exists between deletions (which determine when the rebalancing becomes necessary) and the rebalancing procedure. No algorithm has previously been given for performing a deletion analogously to an insertion, by using merges where the insertion uses splits.

1.3. Symmetric Deletion Approach

Let us consider various implementations of merging a node n with its right neighbor n' in the B-link structure. If we move the data in n' to n (Fig. 3a), there is no path that processes can follow from n' to the data moved out. This may, for example, lead a process to conclude that c is not in the structure.[1] If we move the data in n to n' (Fig. 3b), we will need to access the left neighbor of n to remove n from the linked list. Finding the left neighbor requires either much time or a doubly linked list, which increases complexity and the locking overhead (see [Sh84] for example).

Our novel way of doing merges solves the above problems. We move the data from n' to n, but also direct a pointer called an *outlink* from n' to n (Fig. 4). Any action needing data that was in n', and still expecting to find it there, can access n', see that it is empty, and follow the outlink to n.

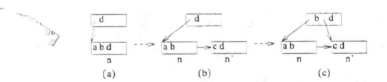

Figure 2. A split in the B-link structure.

Processes looking for d in n in (b) or (c) can proceed to n' and still find d.

---

[1] This illustrates a general principle that data should only move to the right in the singly-linked B-link structure.

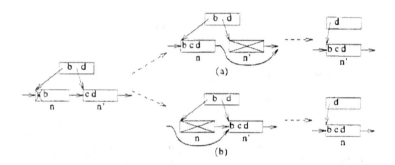

Figure 3. Two possibilities in merging n and n'.

Unfortunately, (a) is incorrect and (b) is inconvenient.

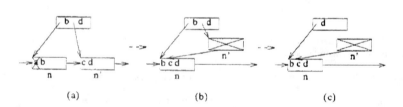

Figure 4. Merging in the B-link structure.

Processes looking for d in n' in (b) and (c) can proceed to n and still find d.

2. Our Algorithm

The following section fleshes out the short description of our algorithm given above. Please refer to Fig. 5 for an example of a B-link structure and to the appendix for a pseudo-Pascal program implementing the algorithm.

2.1. Some Definitions

Each internal node consists of a rightlink, and a sequence $(p_1, s_1, p_2, s_2, \cdots, p_q, s_q)$, where each p is called a *downlink* and s_i a *separator*. A downlink is a pointer to a (possibly

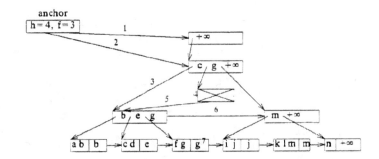

Figure 5. A possible legal state of a B-link structure.
1) The *top* pointer. 2) The *fast* pointer. 3) A downlink. 4) An empty node. 5) An outlink. 6) A rightlink. 7) A leaf's only separator.

null) chain of empty nodes terminating in a non-empty node on the level below. A separator is a value from the same domain as keys, but is not a key itself. For $1 \leq i < q$, $s_i < s_{i-1}$. Each leaf consists of a rightlink, a sequence of keys $(v_1, v_2, \cdots, v_q)$, and a single separator s, where $v_i < v_{i-1} \leq s$ for $1 \leq i < q$.[2]

Let us define:

rightsep(d), where d is a downlink: the separator on the immediate right of d. Note that it is always in the same node as d.

rightsep(n), where n is a non-empty node: the rightmost (largest) separator in n.

leftsep(d), where d is a downlink in node n: the separator to the immediate left of d. If d is the leftmost downlink in n, *leftsep(d)* is the rightmost separator in the left neighbor of n. If n is the leftmost node on its level, *leftsep(d)* is considered to be $-\infty$.

leftsep(n), where n is a non-empty node and d is the leftmost downlink in it: *leftsep(n) = leftsep(d)*. If n is an empty node whose outlink points to n', let us define *leftsep(n) = leftsep(n')*.

coverset(n), where n is a non-empty node: $\{x \mid leftsep(n) < x \leq rightsep(n)\}$.

2.2. Locks

The locking model used in [LY81] assumed that an entire node could be read or written in one indivisible operation. This assumption enabled the [LY81] algorithm to let reader

---
[2] For the sake of simplicity, we have restricted the structure to holding only unique keys.

processes access nodes without first locking them in any way Since the atomicity of node reads and writes is not a reasonable assumption in some environments (such as when the structure is in primary memory), and in order to make comparisons to other algorithms easier, we use a more general locking scheme similar to the one in [BS77] It is a simple matter to convert our algorithm to fit the [LY81] model, as should be done when atomic node accesses are provided by the hardware

Two kinds of locks are used a read lock and a write lock Many processes may hold read locks on a node at the same time, but a write lock demands exclusive access to a node, allowing no other locks A process must hold at least a read lock to perform a non-modifying operation on a node, a process must hold a write lock to modify a node Deadlock is not possible in our algorithm, so lock managers of different nodes may be co-independent

Each action holds no more than one read lock at a time during its descent, an insertion holds no more than one write lock at a time during its ascent, and a deletion needs no more than two write locks at a time during its ascent

2.3 The Locate Phase and the Decisive Operation

The first phase of all three actions (search(v), insert(v), and delete(v)) is called the *locate phase*, as its function is to locate and place a lock on a leaf n such that $v \in coverset(n)$ That is, n should be the node where v should be if it is anywhere in the structure The locate phases of all three actions are identical, except that the last two must leave a write lock on n, not a read lock

The locate phase does not use lock coupling On the path to the appropriate leaf, the locate phase locks each node, determines which link to follow out and unlocks the node It does not, however, unlock the final leaf Each node-to-node transition usually follows a downlink, but follows a rightlink or outlink if necessary Write locks are used on the leaf level if the calling action is insert or delete, and read locks elsewhere

Once the locate phase completes, an action performs its *decisive operation* which consists of checking for v (for search), adding v (for insert), or removing v (for delete) in the locked leaf

2.4 Restructuring

Adding or removing information in a node may leave it too crowded or too sparse As in sequential B-tree algorithms, this is rectified by *normalizing* the node i e splitting it into two nodes or merging it with its neighbor [1] Since a split or a merge involves inserting or removing the pointer from the parent node to the node being created or vacated, the normalization of a node may make further splits and merges higher in the structure necessary Thus the tree is restructured by an ascent from the leaf accessed by the decisive operation

2.4 1. Two-Phase Splits and Merges

In order to avoid having to lock the parent (and possibly more distant ancestors) of a node while normalizing it, our algorithm performs splits and merges in two stages, as shown in figures 2 and 4 The first stage (fig 2b and 4b), which does not involve the parent of the node being split or merged, is called the *half-split* or the *half-merge* respectively After a half-split, n and n' may still be regarded as one logical node The half-merge may be regarded as logically merging n and n', and re-directing to n the downlink to n' A downlink

---

[1] Merging two nodes into one may result in a node that is too crowded Thus a merge may be followed by

is said to *trans-point* to the first non-empty node on the path it starts, thus the downlink to n' trans-points to n. After the completion of a half-split or a half-merge, all locks are released

The second stage (Fig 2c and 4c) is called the *add-link* or *remove link* This operation takes place on an "appropriate" node on the level above the node r which has just been half-split or half-merged In a sequential computation the appropriate node is always the (unique) parent of the node involved. This can not work in our algorithm because a node may (rarely) have several or no parents The purpose of a downlink — to channel to its target the locates of keys in the target's *coverset* — suggests a better criterion Let s be the rightmost separator in the left node after a half-split, or the rightmost separator in the left node before a half-merge The (unique) appropriate "parent" node p is the one for which s ∈ coverset(p)

An add-link consists of inserting s and the downlink to the new node into the sequence in p, with s on the immediate left of the downlink A remove-link consists of removing s and the downlink to the empty node from p, with s on the immediate left of the downlink[4]

Normally, finding the node with the right *coverset* for the add-link or remove-link is done as in [LY81], by reaccessing the last node visited during the locate phase on the level above the current node Sometimes (e g when the locate phase had started at or below the current level) this is not possible, and the node must be found by a new descent from the top

2.4.2. Changing the Height of the Structure

If an ascent can not proceed because the current level is the top level, a new top level is created (by the *grow* operation) Safely removing unneeded top levels turns out to be too difficult However, descending through them for every action is unacceptable Thus, instead of actually removing them, we simply avoid searching through them A continuously running process called the *critic* keeps track of (a good guess at) the highest level (called *fast*) containing more than one downlink The critic uses only read locks in getting this information Pointers to the leftmost nodes of both the fast level and the top level are kept in the *anchor*, whose address is known to all processes If need be, the critic modifies the fast pointer The locate phase of all actions begins at the leftmost node on the fast level (The locate phase will complete at the correct leaf regardless of the precise height of the fast level because it is in fact safe to start a locate at any level) The levels above the fast level remain unused until they are needed again when the fast level contains too many (e g more than 2) non-empty nodes Since these levels normally consist of only one node and number no more than the maximum (over time) height of the structure they take up little space

2 5. Freeing Empty Nodes

Once all pointers to an empty node are removed the node becomes a candidate for re-use[5] However, it can not then be freed immediately since processes may exist that obtained

---

a split this is our version of rotation

[4] There are special cases in performing the remove-link If s is the rightmost separator in p then the downlink is leftmost in the right neighbor of p The simplest solution is to merge p and its neighbor prior to the remove link and to normalize afterwards In rare circumstances s and the downlink will not be in p at all in which case we reissue the remove link later We know that it will eventually succeed because the empty node was originally created by a half split, and all half-splits are eventually followed by the appropriate add-links The remove link is this add link's counterpart and was delayed because the add link was slow in completing Similarly an add-link has to be tried again if s is already present in p this happens only when this older s s remove link has been slow

[5] A counter of the number of outlinks pointing to the empty node must be used Furthermore just the absence of a downlink to the node is insufficient as the add link may simply be very slow and still active The

some pointer to it before the pointer was removed If such a process were to access the node once it was re-inserted into the structure, the process might be misdirected or damage the node

Two approaches are possible One, introduced in [KL80] and called the *drain technique*, is to delay freeing the empty node until the termination of all processes whose locate phase began when pointers to the node still existed The other approach is to store information in every node (its height and *leftsep*) such that a process can recognize that it has become lost and recover by going back to the previous node it accessed or to the anchor The appendix does not contain code for either approach, but freeing empty nodes by the drain technique is transparent to the rest of the algorithm and can be added on as a separate module

3. Correctness

We verify the algorithm by applying the ideas of semantically-based concurrency control to the dictionary abstract data type [GoSh85] (see also [FC84]) Kung and Papadimitriou [KP79] established the theory of a semantic approach to concurrency control by showing the relationship between the information available to a concurrency control algorithm and the achievable concurrency Our algorithm (like Lehman and Yao's) is not serializable in the usual syntactic sense (i e in the unrestricted reads and writes model), thus the more general semantic approach is necessary

The *state* of the dictionary abstract data type (ADT) is a set of keys The set of all potential keys is called *KeySpace* and we assume it contains the values $+\infty$ and $-\infty$ which are, respectively, larger and smaller than all others The dictionary ADT supports the dictionary actions *search(k)*, *insert(k)*, and *delete(k)*, which map a dictionary state s to a dictionary state s' and return value v For search(k), $s' = s$ and $v = (k \in s)$, for insert(k), $s' = s \cup \{k\}$ and $v = (k \notin s)$, for delete(k), $s' = s - \{k\}$ and $v = (k \in s)$ [6] A sequence of dictionary actions $a_1, a_2, \ldots, a_k$ maps an initial dictionary state s_0 to a final state s_k and a sequence of return values $v_1, v_2, \ldots, v_k$, where each a maps s_{i-1} to s_i and returns v It is the job of our algorithm to implement dictionary actions correctly when they execute concurrently.

A search structure state (a set of nodes and edges with associated information) implements a dictionary state For each node n in the search structure, *contents(n)* is the set of keys in the node The dictionary state implemented by a set of nodes N is $\bigcup_{n \in N} contents(n)$ Operations on the search structure state are called *search structure operations* For our algorithm, these are operations like half-split, add-link, and check-key Search structure operations map a search structure state to another search structure state and a return value Each dictionary action is implemented as a program of search structure operations

Let a set of dictionary actions A be run concurrently [7] The resulting computation C consists of an initial search structure state s, a final search structure state s', the set of dictionary actions A, the set of executed search structure operations O, a function *parent* from the members of O to A, a function r from the members of $A \cup O$ to their return values, and a partial order $<$ on the members of $A \cup O$ (where $a < a'$ means a completes before a' begins, and $\neg(a < a' \lor a' < a)$ means a and a' were at least partially concurrent) We assume that

---

correct condition is the completion of the remove link

[6] For simplicity, we omit a formal description of the handling of records usually associated with keys

[7] The following definitions are actually applicable to two level computations on any abstract data type Simply replace the words dictionary' with top level" and search structure with bottom level

$$\forall_{a\,a\,\cdot\,\lambda}\quad a<a' \Rightarrow (\quad \forall_{o\,o\,\in O}\; (parent(o)=u \wedge parent(o')=a') \Rightarrow o<o')$$

Definition A computation C with s, s', A, r and $<$ defined as above is *serializable* if the members of A can be arranged in a sequence A' such that A' maps the dictionary state represented by s to the dictionary state represented by s', produces the same return vales for the members of A as does r, and does not conflict with $<$ An algorithm is serializable if all computations it produces are serializable

In our algorithm, locking ensures that only non-modifying operations are allowed to access the same node concurrently This implies that each operation is *atomic*, i e the final state and return values of every computation are as if the operations occurred in a sequence, i e a total ordering extending the actual partial ordering This makes it possible to think of each operation as occurring alone and starting and finishing in a well-defined search structure state, which facilitates reasoning about computations

We now develop several results needed to show that our algorithm is serializable

Definition A B-link structure state is *legal* if
LS1) There exist values h and f, $1 \le f \le h$, such that the non-empty nodes form h linked lists through the rightlinks, where all leaf nodes are on list 1, all downlinks from list i trans-point to nodes on list $i-1$, and the anchor contains pointers to the source nodes of lists h and f, as well as the values h and f
LS2) The separators on each list form a strictly ascending sequence terminating in $-\infty$ The keys in each leaf are in the leaf's *coverset*
LS3) If a downlink d points to node n, $leftsep(d) \ge leftsep(n)$ (This is the minimal condition sufficient for safe descents If $leftsep(n)$ were greater than $leftsep(d)$, actions could follow d to n while the data they seek lies to the left of n)

B-link Proposition For all operations mapping legal search structure state s to state s', for all nodes n present in s, if $leftsep(n)=l$ in s and $leftsep(n)=l'$ in s', then $l' \le l$

Proof This follows directly from the definitions of the operations, as the reader can easily verify For example, an add-link can not violate the proposition because it never adds the largest separator to a node, thus not changing any *leftsep*s As another example, a half-merge decreases the *leftsep* of the node it empties (by the definition of *leftsep* for empty nodes) This proposition is a formal re-statement of the rule that information never moves to the left in the B-link structure
[]

Lemma 1 If d is one of the operations check-key(x,n), add-key(x,n), or remove-key(x,n) in computation C of our algorithm (i e d is a decisive operation), and C starts in a legal state and all operations up to d finish in a legal state, then n is a leaf and $x \in coverset(n)$ in the search structure state preceding d

Proof It is the locate phase that finds and locks the node on which the decisive operation acts We note that the height of a node is never changed by any operation that maintains a legal state The locate phase starts out at the leftmost node n at some height f Since f is read from the anchor and the heights of all nodes remain the same, the locate phase can correctly calculate the height of any node it accesses Furthermore, since it starts at the leftmost node, $leftsep(n) = -\infty < x$ The next node to be visited, m, is chosen such that $leftsep(m)$ is as large as possible without violating $x > leftsep(m)$ (LS1, LS2 and LS3 guarantee that this decision can be safely based on the values of the separators in n) By the B-link Proposition, $leftsep(m)$ can only get smaller with time Thus, when a lock is placed on m, $x > leftsep(m)$ still Since m now becomes the new value of n, $x > leftsep(n)$ remains invariant throughout the locate phase The locate phase terminates only when the leaf level is reached and $x \leq rightsep(n)$, which, together with the invariant, implies $x \in coverset(n)$[8]

[]

Lemma 2 If o is one of the operations add-link(s,c,p) or remove-link(s,c,p) in a computation C of our algorithm, and C starts in a legal state and all operations up to o finish in a legal state, then a pointer to c trans-points to a node on the level below p and $s \in coverset(p)$ in the search structure state preceding o

Proof The argument that this condition is met (by the locate-internal subroutine) is similar to the proof of Lemma 1 The condition that c is at height one below p is met because node height does not change and is calculated correctly As for the *coverset* condition, it is satisfied as it was for the decisive operations, except that locate-internal does not always start at a leftmost node However, when it doesn't, the *leftsep* of the starting node is known to have been smaller than s at some previous time By the B-link Proposition, the *leftsep* can only have gotten smaller since that time and the invariant condition is met

[]

Lemma 3 If o is an operation in a computation C of our algorithm, and C starts in a legal state and all operations prior to o finish in a legal state, then o will finish in a legal state

Proof Clearly, no operations other than those in Lemmas 1 and 2 are capable of mapping a legal state to an illegal one By the lemmas, these remaining operations must satisfy their respective height and *coverset* conditions, which implies that they too can not violate LS1 or LS2

It remains to consider LS3 The only operations that could conceivably change the *leftsep* of any existing downlink are add-link and remove-link, which add and remove separators as well as downlinks However, since the separator added or removed is the one on the left of the downlink added or removed, these operations do not change the *leftsep* of any existing downlink By the B-link Proposition, the *leftsep* of any existing node can only get smaller Thus, LS3 can only conceivably be violated in the introduction of a new node or downlink Newly created nodes do not have downlinks pointing to them, so LS3 is not violated Thus, a problem can only arise if for some add-link(s,c,p), $s < leftsep(c)$ However, we note that an add link(s,c,p) is done only after a half-split(c',c) which results in $s = rightsep(c') = leftsep(c)$ By the B-link Proposition, $leftsep(c)$ can only get larger between the half-split and the add-link, thus LS3 is never violated

[]

Theorem All computations produced by our algorithm that begin in a legal state are serializable

Proof All keys reside in non-empty leaves By LS1 and LS2, the *coverset*s of the non-empty leaves in a legal B-link structure partition *KeySpace* By LS2, the keys in each leaf are in that leaf's *coverset* Thus, for every key and every legal B-link structure there is exactly one node (i e the leaf into whose *coverset* the key falls) that could possibly contain the key

By induction on Lemma 3 we have that the algorithm maintains a legal state throughout a computation if it begins in one Then Lemma 1 shows that for every decisive operation on key x in node n, $x \in coverset(n)$ and n is a leaf, which implies that $x \in \bigcup_{m - N - n}contents(m)$ in the state preceding the operation

---

[8] Note that there is no guarantee hat the locate phase will termina e For example a slow search could be indefinitely delayed by continuous half splits on a node it needs to access Thus the algorithm is susceptible to livelock

By inspection of each action only the decisive operation is capable of changing the set of keys in the structure and only the decisive operation affects the return value of the action Furthermore, the definition of each decisive operation is identical to the definition of its parent action except that it acts on a single node instead of the entire structure Since we know that this node is the only one that contains any information pertinent to the action by Lemma 1, each decisive operation correctly performs all of the work of its parent action Thus the sequence of actions given by the sequence of their decisive operations would produce the same final set of keys and the same return values as the actual computation
[]

It should be pointed out that although the algorithm is susceptible to livelock, it is free from deadlock on lock resources because of a well-ordering on the acquisition of locks Whenever a process holds two locks simultaneously, the two nodes are always adjacent at the same height, and the lock on the node on the left is always obtained first

4. Performance Simulation

In order to compare the actual performance of the B-link algorithm to that of previous algorithms supporting insertion and deletion, we have run the algorithms under various conditions in a simulated concurrent environment provided by the language Concurrent Euclid [Ho83]

4.1. Model of Concurrency

We compare speed-up, i e the ratio of the time it takes one processor to do a given amount of work to the time it takes n processors to do that same amount of work Ideally, the speed-up achieved by n processors is n, but actual concurrent algorithms achieve lower speed-ups due to interference through waiting for locks

In our simulation, the processors receive a common list of actions constituting the work to be done Each processor removes an action from the list, executes it, gets another, etc When the last action completes, the current simulated time is noted

The only activity which takes time in the simulation is reading and writing nodes (An action reads in a node after obtaining a lock on it, and possibly writes it out before releasing the lock) However, a process may be delayed while waiting for a lock held by another process which itself is reading or writing a node

A node access (a read or a write) always takes exactly one time unit If n processors issue n node accesses at the same time, all n accesses complete at the same time, after one time unit Thus, node accesses are simulated to occur completely in parallel Studies ([SM85], [FJS85]) have shown that when several physical processors contend for a single relatively slow storage device capable of handling only one request at a time, the storage device can become a bottleneck and prevent the realization of most of the potential concurrency Thus, our model of complete non-interference at the physical storage access level is most applicable to systems where data is distributed over many modules (such as disk drives) capable of operating in parallel Our results are actually quite close to those obtained in [SM85] for the Lehman and Yao algorithm in a simulated system containing as many storage devices as processors

4.2. Parameters of the Simulation

For each simulation run, the values of the keys in the initial B-tree and of the arguments of the job actions were randomly and uniformly distributed over *KeySpace* The values picked for insertions and deletions were guaranteed not to be redundant

The parameters that could be varied in comparing the algorithms were the number of keys in the initial B-tree, the order of the tree (i e the out-degree of nodes), and the proportion of searches, insertions, and deletions in the actions Since searches never interfere with each other and in the absence of updates achieve perfect speed-up, including them in the simulation only masks the amount of interference produced by updates, the source of the problem Thus, we excluded searches from the actions performed and simulated a ' mix" where half the actions were insertions and half deletions

Besides the B-link algorithm, we simulated four other algorithms Two of these, algorithm 1 of [BS77] and the [MR85] algorithm, use lock-coupling (and, for [BS77], subtree-locking) with exclusive locks from the top of the tree Algorithm 2 of [BS77] strives to reduce subtree-locking and exclusive locking high up in the tree by having updates first perform an optimistic search-like descent using read-lock lock-coupling We also simulated our own version of the [MR85] algorithm which uses an optimistic descent like algorithm 2 of [BS77] and re-descends by the [MR85] method

We did not simulate algorithm 3 of [BS77] (as well as an analogous version of the [MR85] algorithm) because it differs from algorithm 1 only in the partial substitution of exclusive locks by writer-exclusion (or alpha) locks Although this substitution can improve performance by allowing read-locks to be held on alpha-locked nodes, only searches use read-locks in this algorithm (alpha locking and optimistic descents can not co-exist because of the possibility of deadlock) Since our simulation excludes searches, its results would not be changed by the substitution

4 3 Results of the Simulation

Figure 6 shows the speed-ups achieved by the various algorithms in an initial B-tree containing 800 keys with each node having between 7 and 13 children or keys Under these conditions (and all other update-intensive conditions we have tried), the algorithms that place exclusive locks from the root during the initial descent achieve a maximal speed-up of only about two It is clear that the algorithms which start placing exclusive locks from the top of the tree are incapable of achieving more than nominal concurrency, in this example producing a maximal speed-up of 2 1 Optimistic descents improve this result significantly with our version of [MR85] achieving a maximal speedup of around 10 before ceasing to get better performance with larger numbers of processors The somewhat worse performance of the second [BS77] algorithm must be attributed to the use of subtree-locking by the latter The B-link algorithm, however, gets a speed-up of 26 5 with 40 processors and would probably get a better speedup with a larger number of processors (compiler constraints prevented simulation of more than 40)

Varying the parameters to the simulation brings some significant changes to the results When the algorithms were run on a 7-13 tree with 100 keys, the small size of the tree (only two levels) increased the interference and thus decreased the performance of all algorithms The B-link algorithm was still best, but achieved a maximal speedup of only 7 This figure increased to 14 with an initial tree size of 200 keys

More significantly, increasing the size of leaves greatly improves the performance of the algorithms employing optimistic descent This is because splits and merges become less frequent, increasing the probability that the optimistic descent will succeed In a 20-39 tree with 10000 keys (roughly three levels), our version of the [MR85] algorithm achieves only slightly lower speed-ups than the B-link algorithm, as shown in Fig 7 The second [BS77] algorithm lags somewhat behind these two, probably because of its subtree-locking

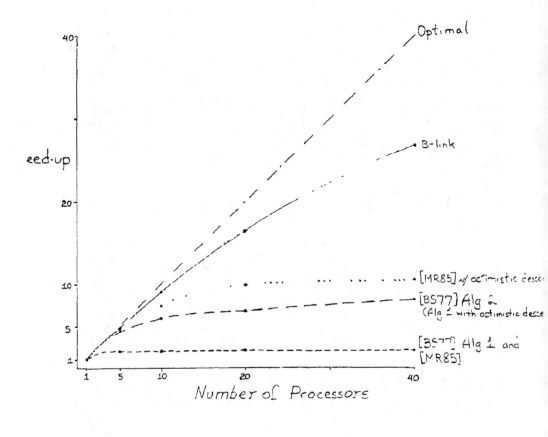

Figure 6

Simulation of insertions and deletions
in a 7-13 B-tree containing 800 keys

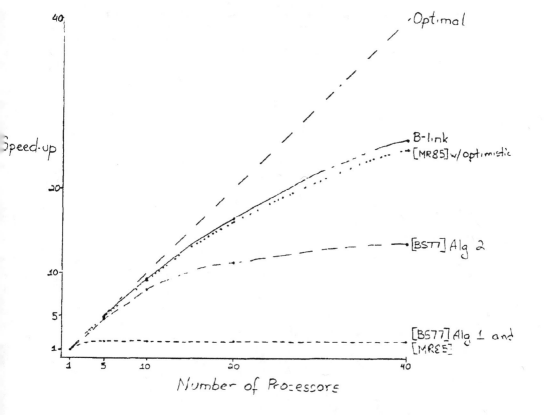

Figure 7

Simulation of insertions and deletions
in a 20-39 B-tree containing 10,000 keys

5 Conclusion

We have presented a deadlock-free concurrent B-tree algorithm supporting search, insert, and delete which employs less locking than any other algorithm that treats deletions symmetrically with insertions To verify the algorithm, we use a correctness criterion based on the semantics of the dictionary ADT This criterion permits executions that would not be considered correct in the model of uninterpreted reads and writes Simulation has shown that the algorithm is capable of achieving much better concurrency than the earliest (and still most widely used) concurrent B-tree algorithms that employ much exclusive locking high in the tree It performs at least as well as other algorithms which also try to avoid using exclusive locks, and performs significantly better than these in trees containing a small number (e g under 20) of keys in each leaf

APPENDIX

The following implementation of our algorithm is written mostly in Pascal Two non-Pascal constructs are used· the use of records as return values of functions, and the use of *spawn* as a way of invoking processes   A value of a record type is represented as $(f_1, \cdots f_n)$ The semantics of *spawn*(procedure call) are to create a process to perform the procedure call   No further communication takes place between the spawner and the process spawned, and the two execute at the same time and terminate independently   The parameters to this procedure must be entirely call-by-value, in particular, the descent stacks must actually be copied[9] No requirement is placed on the length of time that may pass before the spawned process becomes active

The ascent is performed by creating a new process for each add-link or remove-link to be done   We choose this method for simplicity of presentation, although it would be more efficient to have a single process handle all operations on a given level   If the drain technique is used, the starting time of each spawned process must be recorded as the starting time of the parent process

```
type
  locktype = (readlock, writelock),
  nodeptr = ^node,
  height = 1   maxint
  task = (add, remove),

var
  anchor record
    fast. nodeptr, fastheight height,
    top nodeptr, topheight· height,
  end,



function search(v value) boolean,
  var
    n nodeptr,
    descent stack,
  begin
    n = locate-leaf(v, readlock, descent), {v ∈ coverset(n), n read-locked}
    search = check-key(v, n), {decisive}
    unlock(n, readlock)
  end,
```

---

[9] This restriction may be lifted for those parameters that will not be accessed again by the spawner (or passed to another spawned process)

```
function insert(v value) boolean,
  var
    n nodeptr,
    descent stack,
  begin
    n = locate-leaf(v, writelock, descent), {v ∈ coverset(n), n write-locked}
    insert = add-key(v, n), {decisive}
    normalize(n, descent, 1),
    unlock(n, writelock)
  end,
```

```
function delete(v value) boolean,
  var
    n nodeptr,
    descent stack,
  begin
    n = locate-leaf(v, writelock, descent), {v ∈ coverset(n), n write locked}
    delete - remove-key(v, n), {decisive}
    normalize(n, descent, 1),
    unlock(n, writelock)
  end,
```

```
function locate-leaf(v· value; lastlock  locktype, var descent  stack)  nodeptr,
  { locate-leaf descends from the anchor to the leaf whose coverset
    includes v, places a lock of kind specified in lastlock on that leaf,
    and returns a pointer to it  It records its path in the stack descent. }
  var
    n,m  nodeptr,
    h,enterheight  height,
    ubleftsep  value,
    { ubleftsep stands for "upper bound on the leftsep of the current node"
      This value is recorded for each node on the descent stack so that
      an ascending process can tell if it's too far to the right }
  begin
    lock-anchor(readlock),
    n = anchor fast, enterheight = anchor fastheight, ubleftsep = -∞,
    unlock-anchor(readlock),
    set-to empty(descent),
    for h = enterheight downto 2 do begin { v > leftsep(n)}
      move-right(v, n, ubleftsep, readlock), { v ∈ coverset(n) }
      push(n, ubleftsep, descent)
      (m, ubleftsep) = find(v, n, ubleftsep), { v > leftsep(m) }
      unlock(n, readlock),
      n = m
    end,
    move-right(v, n, ubleftsep, lastlock), {v ∈ coverset(n) }
    locate-leaf = n
  end,



procedure move-right(v value var n  nodeptr, var ubleftsep  value, rw  locktype),
  { move-right scans along a level starting with node n until it
    comes to a node into whose coverset v falls (trivially, n itself)
    It assumes that no lock is held on n initially, and leaves a lock
    of the kind specified in rw on the final node }
  var
    m  nodeptr,
  begin {assume v > leftsep(n)}
    lock(n, rw),
    while empty(n) or (rightsep(n) < v) do begin { v > leftsep(n) }
      if empty(n) then m = outlink(n) { v > leftsep(n) = leftsep(m) }
      else begin
        m = rightlink(n), {v > rightsep(n) = leftsep(m) }
        ubleftsep = rightsep(n),
      end,
      unlock(n, rw),
      lock(m, rw)
      n = m,
    end,
  end,
```

```
procedure normalize(n  nodeptr, descent  stack, atheight  height),
  { normalize makes sure that node n is not too crowded
    or sparse by performing a split or merge as needed
    A split may be necessary after a merge  n is assumed to be write-locked
    descent and atheight are needed to ascend to the level above to
    complete a split or merge  }
  var
    sib, newsib· nodeptr,
    sep, newsep  value,
  begin
    if too-sparse(n) and (rightlink(n) <> nil) then begin
      sib  =  rightlink(n),
      lock(sib, writelock),
      sep  =  half-merge(n, sib),
      unlock(sib, writelock),
      spawn(ascend(remove, sep, sib, atheight+1, descent))
    end,
    if too-crowded(n) then begin
      allocate-node(newsib),
      newsep  =  half-split(n, newsib),
      spawn(ascend(add, newsep, newsib, atheight-1, descent))
    end
  end,

procedure ascend(t task, sep  value, child  nodeptr, toheight  height, descent· stack),
  { adds or removes separator sep and downlink to child at height toheight,
    using the descent stack to ascend to it }
  var
    n  nodeptr,
    ubleftsep  value,
  begin
    n  =  locate-internal(sep, toheight, descent)
    while not add-or-remove-link(t, sep, child, n, toheight, descent) do begin
      { wait and try again  very rare }
      unlock(n, writelock),
      delay, { sep > leftsep (n) }
      move-right(sep, n, ubleftsep, writelock) { sep ∈ coverset(n) }
    end,
    normalize(n, descent, toheight),
    unlock(n, writelock)
  end,
```

```
function add-or-remove-link(t task, sep value, child nodeptr,
  n nodeptr, atheight height, descent stack) boolean,
  { tries to add or removes sep and downlink to child from
    node n and returns true if succeeded if removing,
    and sep is rightmost in n, merges n with its right neighbor first
    (if the resulting node is too large, it will be split by the
    upcoming normalization ) A solution that avoids this merge exists,
    but we present this for the sake of simplicity }
  var
    sib nodeptr,
    newsep value,
  begin
    if t=add then add-or-remove-link = add-link(sep, child, n)
    else begin {t=remove}
      if rightsep(n) = sep then begin
        { the downlink to be removed is in n's right neighbor }
        sib = rightlink(n), {rightsep(n) = sep < -∞ thus rightlink(n)<>nil}
        lock(sib, writelock),
        newsep = half-merge(n, sib), {newsep = sep}
        unlock(sib, writelock),
        spawn(ascend(remove, newsep, sib, atheight+1, descent))
      end,
      add-or-remove-link = remove-link(sep, child, n)
    end
  end,
```

```
function locate-internal(v  value, toheight  height, var descent  stack)  nodeptr,
  { a modified locate phase, instead of finding a leaf whose coverset includes
    v, finds a node at height toheight whose coverset includes v
    if possible, uses the descent stack (whose top points at toheight) }
  var
    n, m, newroot  nodeptr,
    h, enterheight  height,
    ubleftsep  value,
  begin
    if empty-stack(descent) then ubleftsep  =  - ∞   { force new descent }
    else pop(n, ubleftsep, descent),
    if v < = ubleftsep then begin
      { a new descent from the top must be made}
      lock-anchor(readlock),
      if anchor topheight < toheight then begin
        unlock-anchor(readlock), lock-anchor(writelock),
        if anchor topheight < toheight then begin
          allocate-node(newroot),
          grow(newroot)
        end,
        unlock-anchor(writelock), lock anchor(readlock)
      end
      if anchor fastheight > = toheight then begin
        n  = anchor fast, enterheight  = anchor fastheight
      end
      else begin
        n  = anchor top, enterheight  = anchor topheight
      end,
      ubleftsep  =  - ∞, { v > leftsep(n) }
      unlock-anchor(readlock),
      set-to-empty(descent),
      for h  = enterheight downto toheight+1 do begin { v > leftsep(n) }
        move-right(v, n, ubleftsep, readlock), { v ∈ coverset(n) }
        push(n, ubleftsep, descent),
        (m, ubleftsep)  = find(v, n, ubleftsep), { v > leftsep(m) }
        unlock(n, readlock),
        n .= m
      end
    end,
    { v > leftsep(n), height of n = toheight }
    move-right(v, n, ubleftsep, writelock), { v ∈ coverset(n) }
    locate-internal  = n
  end,
```

```
procedure critic,
  { the critic runs continuously, its function is to keep the target
    of the fast pointer in the anchor close to the highest level
    containing more than one downlink  }
  var
    n, m  nodeptr,
    h  height,
  begin
    while true do begin
      lock-anchor(readlock),
      n = anchor top, h = anchor topheight,
      unlock-anchor(readlock)
      lock(n, readlock),
      while numberofchildren(n)< = 3 and rightlink(n)=nil and h>1 do begin
        m := leftmostchild(n),
        unlock(n, readlock),
        n = m,
        lock(n, readlock),
        h = h - 1
      end,
      unlock(n, readlock).
      lock-anchor(readlock),
      if anchor fastheight = h then
        unlock-anchor(readlock)
      else begin
        unlock-anchor(readlock),
        lock-anchor(writelock),
        anchor fastheight = h, anchor fast = n,
        unlock-anchor(writelock)
      end,
      delay
    end
  end,
```

{ The search structure operations, arranged alphabetically
Locking ensures that they are atomic }

function add-key(v· value, n nodeptr) boolean,
{ if v is not a key in n then it is added into the sequence
of keys in the leaf n at an appropriate location and true is returned
otherwise, return value is false }

function add-link(s· value, child, parent· nodeptr) boolean
{ The smallest index i in parent such that $s \geq s_i$ is identified
If $s_i = s$, the operation returns false Otherwise, it changes the
sequence in parent to ($,p_i,s,child,s_i,$),
and returns true }

function check-key(v value, n nodeptr) boolean,
{ returns true if v is a key in n, false otherwise }

function empty(n nodeptr) boolean,
{ returns true if n is an empty node }

function find(v value, n nodeptr, ubleftsep value) (nodeptr, value),
{ The smallest s_i in n such that $v \leq s_i$ is identified
If $i > 1$, returns (p_i, s_{i-1}), otherwise $(p_i, ubleftsep)$ }

procedure grow(n nodeptr),
{ n is made an internal node containing only a downlink to the current
target of the anchor's top pointer and the separator $+\infty$ to its right
The anchor's top pointer is then set to point to n, and its height
indicator is incremented }

function half-merge(l, r nodeptr) value,
{ The sequence in r is transferred to the end of the sequence in l
The rightlink of l is directed to the target of the rightlink in r
r is marked empty, its outlink pointing to l If l is a leaf,
its separator is set to the largest key in it
The previous value of the rightmost separator in l is returned }

function half-split(n, new nodeptr) value,
{ The rightlink of new is directed to the target of the rightlink of n
The rightlink of n is directed to new The right half of the sequence in n
is moved to new If n and new are leaves, their separators are
set equal to the largest keys in them The return value is the new
rightmost separator in n }

function leftmostchild(n nodeptr) nodeptr,
{ returns the leftmost downlink in n }

function numberofchildren(n nodeptr) integer,
{ returns number of downlinks in n }

function outlink(n nodeptr) nodeptr,
{ returns the outlink in node n }

function remove-key(v value, n nodeptr) boolean,
    { if v is a key in n, it is removed and true is returned
    otherwise, return value is false }

function remove-link(s value, child, parent nodeptr). boolean,
    { If the sequence in parent includes a separator s on the immediate left
    of a downlink to child, the two are removed and true is returned,
    otherwise the return value is false }

function rightlink(n nodeptr) nodeptr,
    { returns the rightlink in node n }

function rightsep(n nodeptr) value,
    { returns the rightmost separator in n }

function too-crowded(n nodeptr) boolean,
    { returns true if n contains too much information }

function too-sparse(n nodeptr) boolean,
    { returns true if n contains too little information }

REFERENCES

[BS77] BAYER, R , and SCHKOLNIK, M Concurrency of operations on B-trees Acta Inf
9 1 (1977) 1-21

[Ca81] CASANOVA, M A *The concurrency problem for database systems, Lecture Notes in
Computer Science, vol 116, Springer-Verlag 1981*

[Co79] COMER, D The ubiquitous B-tree ACM Computing Surveys 11,2 (June 1979)
121-137

[El80] ELLIS, C Concurrent search and insertion in 2-3 Trees Acta Inf 14,1 (1980) 63-86

[FC84] FORD, R and CALHOUN, J Concurrency control mechanisms and the
serializability of concurrent tree algorithms Proceed of the ACM Symp on the Princ of
Database Syst 1984

[FJS85] FORD, R , JIPPING, M , and SHULTZ, R On the performance of an optimistic
concurrent tree algorithm Technical Report 85-07, Dept of Comp Science, University of
Iowa, 1985

[Ga83] GARCIA-MOLINA, H Using semantic knowledge for transaction processing in a
distributed database ACM Trans on Database Syst 8,2 (1983)

[GoSh85] GOODMAN, N , SHASHA D Semantically-based concurrency control for search
structures ACM SIGACT-SIGMOD Symp on Princ of Database Syst , 1985

[GuSe78] GUIBAS, L , and SEDGEWICK, R A dichromatic framework for balanced trees
Proc 19th Annual Symposium of Foundations of Computer Science, 1978, 8-21

[Ho83] HOLT, R C *Concurrent Euclid, the Unix system and Tunis, Addison-Wesley Reading
Mass , 1983*

[KL80] KUNG, H T , and LEHMAN, P Concurrent Manipulation of Binary search trees
ACM Trans on Database Syst 5,3 (1980) 339-353

[KS83] KEDEM, Z and SILBERSCHATZ, A Locking protocols from exclusive to shared
locks J of the ACM 30,4 (1983) 787-804

[KW82] KWONG, Y S and WOOD, D Method for concurrency in B-trees IEEE Trans
on Software Engineering SE-8,3 (1982) 211-223

[LY81] LEHMAN, P , and YAO, S B Efficient locking for concurrent operations on B-
trees ACM Trans on Database Syst , 6,4 (Dec 1981) 650-670

[MR85] MOND, Y and RAZ, Y Concurrency control in B$^-$-trees databases using
preparatory operations Proceedings of Internat Conf on Very Large DataBases, 1985,
331-334

[OG76] OWICKI, S , and GRIES, D An axiomatic proof technique for parallel programs 1

www.ingramcontent.com/pod-product-compliance
Lightning Source LLC
LaVergne TN
LVHW012201040326
832903LV00003B/41